>>> **e-guidelines 12**

Handheld technologies for mobile learning

Di Dawson

21 De Montfort Street
Leicester
LE1 7GE

Company registration no. 2603322
Charity registration no. 1002775

NIACE has a broad remit to promote lifelong learning opportunities for adults. NIACE works to develop increased participation in education and training, particularly for those who do not have easy access because of class, gender, age, race, language and culture, learning difficulties or disabilities, or insufficient financial resources.

You can find NIACE online at www.niace.org.uk

Cataloguing in Publication Data
A CIP record of this title is available from the British Library

Designed and typeset by Book Production Services, London
Printed and bound in the UK by Latimer Trend
ISBN: 978 1 86201 320 9

Contents

Acknowledgements

Thanks to the people who helped in the production of this book including Alistair McNaught, Dave Sugden, Terry Loane, Eta de Cicco, Sue Southwood, Carl Hilsdon and Peter Lavender.

Thanks also to the following organisations:
> Adit Ltd and Dolphin for permission to use their logos.
> TechDis for permission to use the screenshot on p. 44.
> Microsoft product screenshots reprinted with permission from Microsoft Corporation.

Supporting web pages

In conjunction with the production of this book a set of supporting web pages with additional resources have been provided at **http://www.niace.org.uk/mobiletechnology**

1

Introduction

This e-guideline draws on the experiences and findings of many projects which can be classified as M-learning (mobile learning). It aims to offer you practical ideas on how handheld technologies can be used in education. It is not a technical manual, however there are occasional tips and hints that you should find interesting and helpful.

Mobile technologies offer opportunities to provide new and interesting ways of learning and can motivate and encourage adult learners to achieve success. As these technologies become more widely available in general use, this book sets out to explore how they can be employed as educational tools.

This guide covers a range of mobile technology that can be used inside and outside the classroom, including the use of Personal Digital Assistants (PDAs), mobile phones and digital audio players such as iPods. Distinctions between these devices are now becoming blurred as technologies converge as, for example, with the 'smartphone' which combines PDA and mobile phone features. It focuses on handheld devices but does not include handheld games consoles.

A Personal Digital Assistant (PDA) is a handheld computer

Not everyone learns the same thing in the same way and handheld devices in a classroom can offer learners a variety of experiences and opportunities to learn in different ways. Activities described in this guide show practical ways to use handheld devices in a variety of teaching and learning environments.

The portability of handheld devices means that tutors working away from main educational sites can still use technology in their teaching. Handheld devices require little set-up time and minimal support. For many learner, handheld technology can appear less threatening than a laptop or desktop computer and can allow technology to be used in a more personal way. For young adults, mobile technologies can have particularly broad appeal and engage them in learning where other methods may have failed. Devices have become more powerful in processor, memory and battery life; the functions they possess are more numerous and flexible. Added to this, the cost of technology is falling.

A smartphone combines the features of a mobile phone and a PDA

In the past few years there have been many trials, research projects and developments looking at the use of handheld devices and their application to learning. Trials, reported in JISC's *Innovative Practice Guide*, have considered a number of approaches using handheld devices. Some of the scenarios in the Guide show that the devices have supplemented activities, whereas others indicate that tutors have designed blended activities around the mobile devices. Practitioners are also prompted to consider the following:

> the learner's needs
> where the learning is taking place
> the outcome of the learning

There is a Glossary at the back of the book which provides explanations of words and acronyms.

For further examples of practice and technical information see the website which has been developed to support this book: **http://www.niace.org.uk/mobiletechnology**

A MP3 player will play sound files and can be used to make audio recordings

2

Why use handheld devices in teaching and learning?

This section explores some of the benefits of using handheld devices in teaching and learning. It covers:

> Cost
> Ownership and empowerment
> Anytime, anywhere learning
> Size and portability
> Attitude to technology
> Collaborative learning
> Text messaging
> Privacy

Cost

Many handheld devices are cheaper to purchase than desktop computers or laptops. This can allow an organisation to introduce technology to learners at a lower cost. Handheld devices also require less technical support than computers and laptops. The greatest need for technical support is prior to initial use where synchronisation software and Internet connections need to be set up. Insurance premiums and the cost of providing secure storage should also be considered.

Before purchasing a handheld device, it is important to consider its features, functionality and the rationale for use with learners. (See checklist p. 40) Organisations may wish to consider online purchasing of ICT equipment as selection can be simple, costs lower and delivery faster. Bulk purchase can also show savings.

Ownership and empowerment

Using handheld devices can support learners to take control of their own learning both inside and outside of the classroom. However, where the equipment belongs to the education or training organisation it is important to consider how it will be managed. Before issuing handheld devices to your learners, you should consider your procedures including:

- devising an acceptable use agreement;
- asking learners to sign written agreements regarding use;
- requesting financial deposits (where devices are to be taken off site).

Another option is to use the personal devices that learners bring to the class. Marc Prensky (Prensky, 2001) describes the young student of today as being a 'Digital Native'. Digital Natives have grown up surrounded by computers, video games, digital cameras and video cameras, mobile phones and, of course, the Internet.

Prensky suggests that tutors need to be aware of the familiarity and comfort that young people have with technology and to provide the tools and activities for accessing and engaging with others. As you are likely to have young people in a class it's important to be aware of their familiarity with, and access to, mobile devices.

If learners use their own personal devices then they are likely to be readily available at all times, in their pocket, jacket or handbag. They are also likely to be confident users and may be able to customise the device to personalise their learning.

In practice

'Phones and MP3 players are a portable, affordable, and culturally relevant way for learners to engage in novel situations. Clearly these present new ways to provide students with greater experience, engagement and independence in the classroom but especially in fieldwork and in collaborative groups.'

Alistair McNaught, Senior Adviser, Techdis

Top tips

1. Put the learner's needs first
2. Check the device is appropriate for the learner
3. Check the device is appropriate for the learning situation
4. Consider how using the device will enhance learning and the learning outcome
5. Identify advantages and disadvantages of using learners' own devices or those of the organisation

6 Draw up agreements for use of loan equipment

7 Ensure that you have technical support

Anytime, anywhere learning

Perhaps you've seen someone travelling on a bus or train or walking along a street *'wired for sound'*? You might think it is music which is being listened to but it could be a trendy way of learning on the move. The listener could be reviewing phrases in Spanish or French, listening to a lecture or tutor notes, or watching a presentation or demonstration.

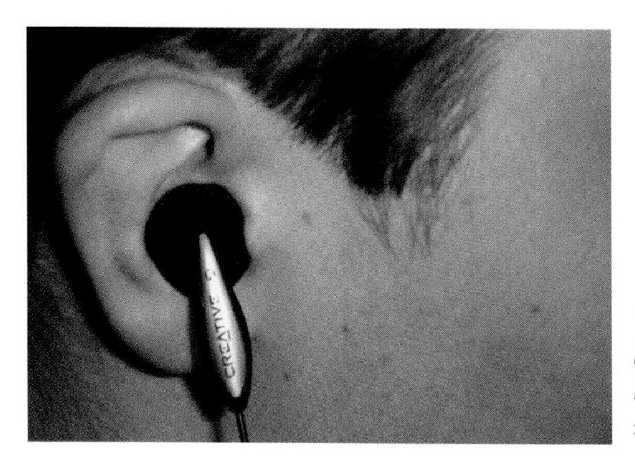

'Wired for sound' - a learner listens and learns while on the move

Epic Group plc's M-learning white paper highlights the potential of mobile devices in reinforcing learning by repetition. The report suggests that

> *repetition is ... the most powerful of learning factors. And by spacing that repetition over time you significantly minimise that forgetting. This may be the primary reason for considering m-learning. At last we have the means to deliver content, participation and regular reinforcement to learners whatever they want, wherever they are and whenever they want it.*
>
> *(Epic Group)*

Learning materials can be devised to be used on mobile devices in 'fieldwork' locations. An example of this was the BBC Springwatch project where participants were asked to go out into the countryside and collect survey data and then use their mobile phones to submit this to the BBC. PDAs can also be used in fieldwork to record observations.

Size and portability

It's much easier to carry and organise mobile devices rather than laptops or desktop computers. A set of 12 PDAs is much lighter and easier to carry into a class in an outreach location. The portable devices offer a further advantage in that they can be used without mains power supply or to access the Internet directly.

Using a stylus with a PDA

Attitude to technology

Some learners might be hesitant when introduced to a computer, keyboard and mouse for the first time, and in some ways this cautious attitude might prevent valuable learning taking place. The PDA, or a mobile phone, can be seen as less-threatening – the user can just touch the screen or use the stylus to navigate around the programmes and screen. Using handheld devices can help overcome barriers to using ICT generally.

For many learners a handheld device can appear friendly and familiar. One learner said: 'It looks like a Gameboy my son has.' It is important to be aware of the learner's attitude towards the use of mobile technology and build on these reactions positively.

Many projects have discovered that the use of handheld devices in learning situations has encouraged individuals who lack confidence in their ability to learn. For young people particularly, their positive attitude towards using mobile technology can be harnessed and can enhance their attitude towards learning.

In practice

'In a Family Learning class two lads were bored so the tutor used the PDA and asked them to write up their experiences. They were immediately happy as it looked street-cred.'

Family Learning tutor, Cornwall AES

Collaborative learning

To include collaborative activities in a learning environment is to offer valuable opportunities for learners to share and build on personal experiences and ideas.

Handheld devices can provide the chance to be involved in problem solving situations, to work collaboratively to create a document or to share sound recordings or images for an end product. Some devices allow for text messages (SMS) to form the basis of a shared communication. The portability of handheld devices meant collaborative learning doesn't need to be confined to the classroom.

Text messaging

Short Messaging Service (SMS) or text messaging has now become a popular way to communicate.

Text messages can be used with learners to:

> provide 1:1 support to individual learners
> broadcast messages to groups of learners
> participate on web-based text message boards
> give swift feedback on a draft or other piece of work
> ask for ideas on a particular topic.

There are many creative ways of using SMS. One way is to encourage learners to 'voice' their opinions by sending a text to each other or to the tutor. Another is to consider the use of text quizzes where a learner sends answers by text to either the tutor or directly to a web area, which then offers immediate feedback. Web-based bulletin boards can incorporate multimedia (MMS) messages with still and moving images, and SMS messages to record a group's collective activity or the individual or group's achievement and progression.

Tutors can use group SMS as a supportive mechanism to advise the learning group about homework, revision or useful ideas and reminders throughout holiday breaks. This type of regular contact helps all learners but particularly those with organisational and memory difficulties.

For information on the use of SMS text see **http://www.text.it/**

Privacy

The portable nature of the devices offers the learner the chance to reflect privately, carry out self-evaluation and capture their thoughts on the device either in text or verbal recording. Handheld devices can be used to take personal notes, review an Individual Learning Plan, Records for Recording Progression or a personal learning diary.

In practice

'Some of my Skills for Life students bring in very personal information – e.g. bills and tax rebates. Using the separate Bluetooth keyboard they can type a letter more privately than at a pc screen then transfer and print it.'

Skills for Life tutor, Cornwall AES

Handheld devices can offer convenient, non threatening access to information technology at a low-cost. Where the application has been chosen to meet an identified educational need then the results can be very effective in improving learning experiences.

3

The potential for each application

In this chapter we will look at a range of applications or functions which are available on handheld technology. We will explore the way in which these can be used to support learning.

Reading text

Text can be read on many mobile devices, providing 'anywhere' access to reading materials. PDAs and Smartphones offer the opportunity to use text-based activities for reading or capturing text. A range of reference material is available for most handheld devices including dictionaries, thesauruses, phrase books, encyclopaedias and e-books.

Text can be read on many mobile devices. This can be used for learning in a number of ways. Here are some examples:

> Read for reference or pleasure a selection of e-books, key texts and handouts relating to a course.

> Privately refer to key materials such as dictionaries and thesauruses where paper versions are not easily available. Online reference materials can be accessed if connection is available.

> Work in pairs to use translators and phrase books for help in modern foreign language or ESOL classes.

- Use sequential skills in referring to databases when off site, e.g. to refer to a wild birds database or foodstuff within a supermarket.

- Follow instructions, for example cooking from a downloaded recipe or one emailed by a friend or tutor.

E-books

An e-book is an electronic (or digital) equivalent of a conventional printed book – it can be displayed on a variety of screens including those on handheld devices. An e-book can be used in order to refer to notes or handouts on the move but also to engage and motivate an unconfident reader who can read a short passage from a book or refer to guidance or review notes.

An e-book can offer much more than a paper book. Text can be highlighted, word searches can be carried out, the built-in dictionary and thesaurus can define words, bookmarks can be placed anywhere within an e-book and notes can be added for later reference. An e-book can also be transferred to another device so a document or e-book can be shared amongst learners, which can encourage peer critique. A further advantage of an e-book is that the text size can be increased for easier reading for learners with visual difficulties.

Learners can also become the creators of e-books by producing a Word document and converting it, using specific software, to e-book format. This can be achieved individually or collaboratively using appropriate text, images, sounds and video clips which can bring an e-book alive and become meaningful.

For more ideas on the use of these devices to support reading or referencing see PDA resources at **http://www.techdis.ac.uk/**

TRY Consider your teaching subject and identify one text-based document or reference material which could be created into an e-book. Ask your learners whether they would find the document useful in a 'handheld size'.

Another method of handling text-based material on a PDA or Smartphone is to save a document in PDF format (Portable Document Format). This can be transferred to a PDA or Smartphone for the learner to read. A PDF document cannot be edited or altered so it is useful for assessment tracking sheets or Individual Learning Plans or any document which doesn't require editing.

Material can be downloaded from the Internet to your computer. The material or programme can then be synchronised, i.e. the files are copied to a PDA or Smartphone.

You can search on the Internet and find speaking dictionaries, phrase books and verb conjugators. Lexisgoo, for example, is a speaking English dictionary which will work on Smartphones that have Pocket PC or similar operating systems. You can even find a choice of medical dictionaries or subject-based encyclopaedias.

TIP

Some e-books offer free trials before purchasing the complete versions and it's sometimes best to check the materials will work well on your device before buying.

A specialised e-book reader device is required to view e-books. You will need to check the device to see whether reading software has been added to it or whether you'll have to download the reader first. For example, Pocket PC users need to look for Microsoft Reader for the reader and a choice of e-books. E-Reader is the software which is available for Palm, Symbian-based phones and Blackberries.

Did you know?

The Sony Reader is a handheld device that provides a 6" screen, a hard drive memory to store approximately 80 e-books and shows a realistic paper background, page turning movements on screen and a 'thumb' indicator to follow the text. It can also play audio files including audio books.

Did you know?

Not all learners can see text on handheld devices. For visually-impaired learners they can choose to purchase software which can be added to a mobile phone so that it 'speaks' text to the user. For example, when Mobile Speak is installed onto a mobile phone it converts text into speech. The speech is heard through the phone's internal speaker or headphones. When the phone rings, the ID number or name is spoken. Similarly, requests for information on the battery status, signal strength and other functions can be voice activated.

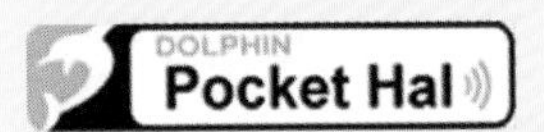

Pocket Hal is a full screen reader created by Dolphin, for the Pocket PC.

Capturing text

Text can be captured on many mobile devices and this facility can be used for learning in a number of ways including:

- Learners can practise note-taking skills and compare results immediately.
- A swift written record can be made of a learning activity even if it is 'off site'.
- Learners can contribute to a collaborative document by passing a device round a group and inviting each group member to add a section.
- Learners can immediately create a blog entry or another e-portfolio contribution at a key point in the course even if there is no PC available.
- Learners can create text on a PDA which is later transferred to other devices like a PC or laptop.

It is important to allow time for learners to familiarise themselves with the device before they use it for a challenging task associated with the course. Fun activities can be devised to allow learners to familiarise themselves with the input method (e.g. stylus or folding keyboard) to give thought to which method will suit each learner.

All phones allow the use of the phone keyboard to input text. A few Smartphones have touch screen facility so you can use a stylus to handwrite notes.

If you need to type a large quantity of text you might find the little onscreen keyboard or phone keyboard tiring to use, so a separate keyboard is useful to have. Some keyboards can be purchased that fold or roll up easily, some can communicate with the device remotely and others offer a cradle to slot your device into.

Many mobile devices allow users to record text in either handwritten or typewritten format. Most PDAs offer three ways to enter text:

1. Handwrite notes
2. Handwrite then convert handwriting to typed text
3. Enter typed text using onscreen or attached keyboard

Apart from capturing text into word-processing or spreadsheet software, a learner can write into a **Note Pad** using handwriting or typed text. This allows, for example, learners to write a personal reflective diary entry immediately and to store the document directly onto a Flash memory card for safekeeping. Alternatively, the online diary entry or blog entry can be uploaded directly to the web if connection to the Internet is available.

Handwriting Tips

For handwriting there is usually a certain hand movement needed to form the letters and words to ensure successful recognition. However, many tutors indicate the value of using the handwriting recognition tool to practise handwriting.

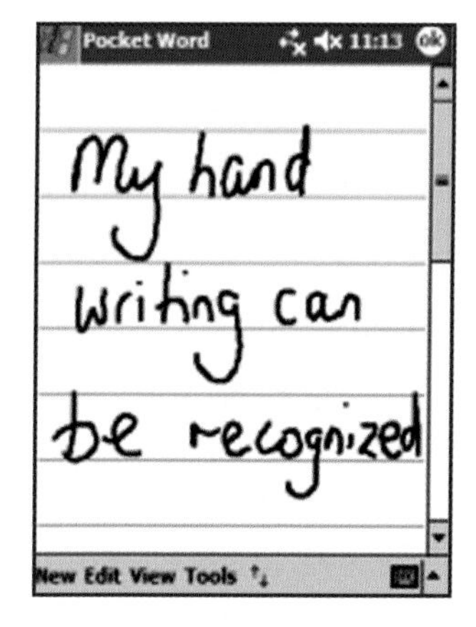

Blogging

A weblog, usually referred to as a 'blog', is like an online diary. These have a very clear application to learning by providing learners with a way of recording the progress of their learning and reflecting upon it. Blogs have also been applied for peer review of work and can be used to practise written language skills.

A website, supporting a blog, displays the written entries to the journal in reverse chronological order. Blogs offer world news, personal reflections and ideas or comments about particular subjects and many include video, images and sounds. Anyone wanting to share information about themselves with others can publish their blog onto the Internet – the most visited blogs are about hobbies, interests and home life. Learners can be encouraged to keep a blog to reflect their learning journey or work in groups to create a shared blog linking to the subject they're studying. A blog can also invite conversations with others by providing the possibility to comment on each post made by the author. Blogs are also a useful and easy way for tutors to publish anything on the web for learners to access.

In practice

'BlogsinHand software was stunning. We had to load the software onto all the PDAs ... it's ideal and works well for the learners to be able to do their blogs on them.'
(http://www.itweaks.com/staticpages/index.php/bihdoc)

Manager, Cornwall Adult Education Centre

Spreadsheets on PDAs

Pocket Excel, a spreadsheet package, has been used with learners in *Skills for Life* classes to collect and count road signs and information.

The survey results were collected and stored in a spreadsheet, however, as Pocket Excel has no graphing capability, an immediate translation from data to graph format was not possible.

Using his computer, the tutor prepared the charts using the data and created images of the graphs, which were then loaded onto the PDA. At the next class the learners could then refer to the chart and review the data collected the previous week.

In practice

'You need to really slow down to write carefully so that it'll recognise your writing. It's great for learners to practise handwriting. It's more difficult though for left-handed users.'

Skills for Life tutor, Cornwall AES

Using sound

It is possible to listen to sound on many mobile devices. This can be used for learning in a number of ways including:

> Listening to instructions of a process or sequence to practise language and listening skills.

> Listening or re-listening to a radio broadcast.

> Listening to podcasts of news, stories and interviews to identify style, genre, use of language or analysis of content.

> Listening to other learners' sound bites or anecdotes about a particular aspect of a course or subject matter or about their own learning.

> Listening to prompts for evaluation so the learner can then record his/her own comments.

> In groups, listening to a mini-lecture for a sound prompt for discussion.

> Listening to audio books for pleasure, inspiration and information. Accompanying transcripts of the sound file help the reading process.

Many people have a preference to learn something new by listening to someone – whether to a mini-lecture, real-life language dialogue or by tuning into a radio programme. Audio format is also ideal for flexibility – why not listen to your notes from a French class rather than looking at them?

The PDA, mobile phone or digital audio player, like an MP3 player, can capture and record sounds as you experience them – at a meeting, in a class discussion or out in the community. On the other hand, the device can become the player so you can listen to previously recorded sound files.

Sound files can also support learners so they can read typed text and at the same time listen to a verbal version of the text. This not only encourages more accurate reading and comprehension but also can help to improve listening skills.

In practice

Cornwall Adult Education Service offer the combination of transcripts and podcasts to *Skills for Life* learners. 'Once you've got the podcast running – they can follow the audio looking at the written word.'

'There is a library of podcasts from the BBC but I've also used the South Yorkshire Police website to download useful podcasts like Internet Fraud. They also have transcripts available.'
(http://www.southyorks.police.uk/podcasts)

Skills for Life tutor, Cornwall AES

A podcast is a sound file (in MP3 or other format) and all you need is a computer to hear this file. However, if you want to listen to the sound file while you're on the move then you need an MP3 player or a device that can play MP3 sound files, i.e. a PDA, mobile phone or iPod.

A sound file usually remains static and unchanged; once you've downloaded and saved it then you can listen to it as many times as you wish. However, a podcast is a little different in that it is a sound file which is replaced and becomes updated when a new 'episode' exists. For example, a radio programme might offer a podcast sound file today which gives you the latest interviews, discussions or ideas from the presenter and today's guests, however, next time you visit the website or plug in your MP3 device into the computer you'll be prompted to download the latest version of the radio programme.

When you search the web and find a suitable podcast you'll be prompted to sign up or 'subscribe' to receive it regularly. Once you've 'subscribed' (usually a free service) the software allows you to search for and download automatically a new 'episode' whenever one is produced. By installing podcast receiver software on your PC you can manage your subscriptions and ensure that you download the latest editions of podcasts. Examples of podcast-receiving software include Juice, Primetime and iTunes.

TIP

News websites, language and travel websites tend to have podcast links for you to catch their latest broadcast. A range of podcasts, including international language podcasts, is available at **http://www.podcast.net.**

Language learners can benefit from listening to native speakers, their tutor and other learners. Try the Learning and Teaching Scotland website for a choice of modern foreign language podcasts. The podcasts offer ideas to consider when creating your own language podcasts. **http://www.ltscotland.org.uk/**

TIP Listen to the BBC's Ouch! It's a Disability Thing podcast – a monthly digest of news and gossip concerning disability and accessibility from radio and TV. **www.bbc.co.uk/ouch/podcast/**

Did you know?

The Podcast for Educators weblog provides ideas and discussion about the use of podcasts in education. It asks 'Is providing educational podcasts via mobile phones the next evolutionary step in podcasting? The advent of 3G and 4G mobile phone networks introduces broadband capability to mobile phones. With established international network infrastructure and an ever-increasing consumer base worldwide, this could really become another growth area for educational podcasting.' **http://www.recap.ltd.uk/podcasting/**

Capturing sound

Sound can be captured on many mobile devices. This can be used for learning in a large number of ways including:

> creating verbal instructions for others in a particular language;

> creating immediate sound bites – recording noises of the street, animals, interviewing people talking in anecdotal language – for later analysis or use in quizzes, presentations etc;

> recording an audio book from a learner's transcript;

> working collaboratively to create a 'sound history' of the local community;

> recording a collaborative story, where students take it in turns to contribute;

> recording and reviewing learners' presentations;

- creating ‘real world’ dialogue and discussion for current or future learners in a subject area;
- working with others to capture pieces of music;
- capturing a verbal reflective diary, especially useful for environments where ICT equipment is less likely to be available, e.g. swimming classes, exercise groups;
- recording evaluative comments regarding the session and course – as an individual, in pairs or as a group;
- recording individual course aims from each learner;
- recording examples of learners’ oral skills, reviewing them as the course progresses;
- recording a discussion for later analysis;
- recording learners’ self-assessment after a session;
- recording a final presentation by learners which draws together the product of their learning (e.g. a podcast);
- creating audio materials by the tutor for use by learners;

Creating your own sound files means that you can capture the tone, use of language and content you wish to achieve.

The BBC and many other websites offer a range of ready-made educational and informative sound files. There is also great benefit for tutors and learners if they create relevant recordings appropriate to the subject area they are studying. Learners can not only gain the technical skills in creating the sound files while they’re practising oral and presentation skills, but also work with others to share the sound creations.

Capturing a discussion can be used as a focal point for creative writing or language development. The verbal notes can be shared easily and have the added function of allowing listeners to hear intonation and inference as the speakers have their say. Language based classes such as Modern Foreign and English Language classes, *Skills for Life* communication activities, discussions and debates can be captured using these devices. The sound file could then be stored on the device, shared with other devices or transferred to a computer to be included in a PowerPoint presentation or Word document.

Handheld devices also offer the opportunity for learners to sit in a quiet place and record their reflective comments in private. For some learners, a verbal recording allows them the chance to be more frank and open than if the thoughts were to be converted into a written passage. Recording a learner talking about their progression and achievement can sometimes be more manageable using a handheld device than pen and paper.

In practice

'We've been experimenting using MP3 sound recording to capture students' evaluations. They're more articulate if you're face-to-face and without the "barrier of paper". We've also captured comments and sounds on a trip to Wales including the Wales Sheep Voice Choir which can be heard at the end of one of the podcasts.

At the moment we're considering using music, which the students like, overlaid with advice and guidance "sound bites" as broadcasted sound throughout a well-used corridor at college – a "wired corridor" where there could be "learning by stealth mode"'.

(http://www.preston.ac.uk/MSICT/index.htm)

TRY

Find the voice recording function on a mobile phone or PDA and record a short message. Listen back to hear the clarity, tone and whether you can hear hiss or background noise.

In practice

Peterborough Adult Learning Service has been using MP3 player/recorders in swimming classes to capture learners' comments before and after sessions in the pool. They also use MP3 players in ballroom dancing classes to record the learners' voices and to capture their progression throughout the classes. In conjunction with a digital camera, instant moments are captured and later reviewed by tutor and learner.

As many learners prefer to learn using auditory methods, Redcar Adult Learning is considering the recording of instructions for basic IT procedures using a visual language to describe procedures such as how to save a document. They will store the sound files on a PDA to offer learners privacy when listening to the instructions.

You can record the sound files using your computer and a microphone and then transfer the files to an MP3 player, mobile phone or PDA. Alternatively, use the device to capture the sounds directly. Many digital audio players can operate voice recording and have an integral microphone to capture voices close by. Some devices have external microphone facilities, which is preferable if you wish to record a larger group of voices.

TRY Use the Internet to find information on ways to record sound and how to insert sound files into documents. Try **http://www.aclearn.net**

Sound recording using Audacity

Audacity is open source software which can be installed at no cost on a PC. Once loaded onto a computer, it can be used to make live audio recordings through a microphone or digitise recordings from cassette tapes or vinyl records (subject to copyright restrictions). This might help language tutors who have many resources in cassette format and might wish to digitise and transfer them to an MP3 player. With Audacity you can import many types of sound files, edit and combine them with other files and recordings. It also allows for fade in and out, pitch change and removes hiss or background noise. The resulting digital files can then be transferred to mobile devices. **http://audacity.sourceforge.net/**

Using images

Images can be seen on many mobile devices. They can be used for learning in a number of ways including:

- contributing to a collaborative visual story of a particular topic – of the local community or about family and friends;
- creating a visual reflective diary or blog capturing visual prompts accompanied by sound or text;
- providing a visual reference before visiting a museum or art gallery;
- creating an index of items such as flowers, weeds or birds;
- providing visual stimulus to encourage creativity in arts, crafts, language and communication activities.

Some people have a preference for learning with visual images and would prefer to see photos and moving images stored on their pocket devices. Most PDAs and many camera mobile phones can support still images as well as short video clips. However, some learners may not respond positively to images so beware of relying on them too heavily.

The quality of video playback and the ability to pause and rewind is a great benefit to learners practising or reviewing a demonstration of a technique or to see an animated group discussion. As devices contain larger memory capacity, so storage of a greater number of images and video clips is possible.

TRY

If you take many photos using a PDA always remember to transfer the images to a computer, laptop or to a removable Flash memory card in case your PDA loses power and you lose precious photos.

Slideshows can be created using specific software such as Microsoft Photostory, which is free to download to a computer. The software allows you to add images, text and sound to the slideshow and finally prompts to save on a computer, Pocket PC device, Smartphone or a portable media device.

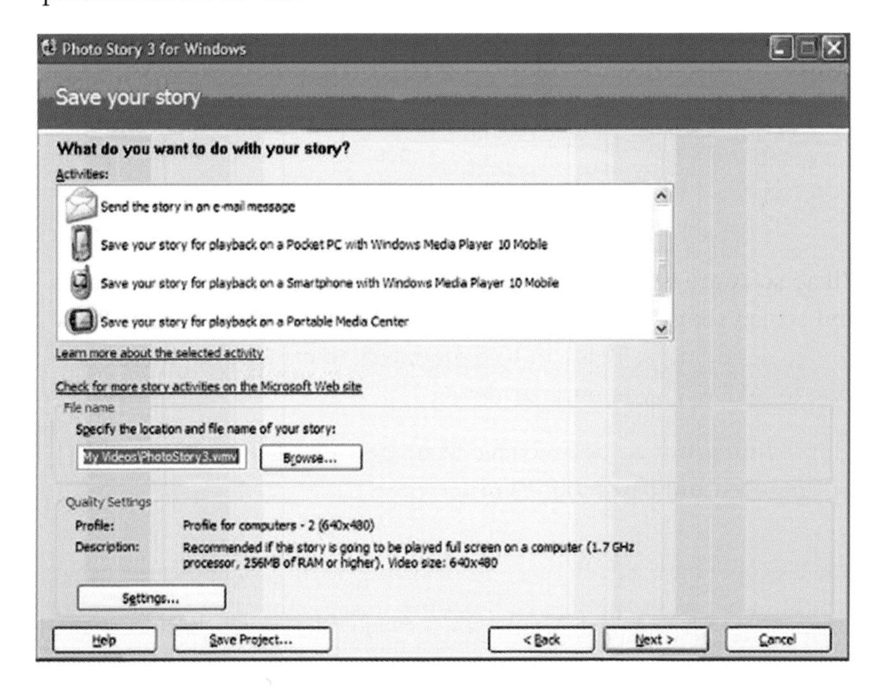

Screen view of Photostory at the point of saving the file

In practice

'Working with families we use the software called Photostory. It's great! We use it for trips out, e.g. a treasure hunt when the kids take the photos along the trail and parents record the sounds. The experience can be captured in a Photostory, which can then be uploaded to the PDA to see the complete presentation.'

Family Learning tutor, Cornwall AES

In practice

Learners at Cornwall Adult Education Service captured images using their PDAs then uploaded the images to a computer to create a Photostory slideshow. Next they converted the slideshow back to the PDA or to their own personal device so that the story became portable and could be shared with friends and family.

In practice

Gloucestershire College of Art and Technology (GLOSCAT) uses presentations prepared by the British Nutrition Foundation for cookery and healthy eating classes. They are freely downloadable from **www.nutrition.org.uk**. GLOSCAT uses devices which run Windows Mobile 5.0, and therefore Mobile Powerpoint shows full animation throughout the presentation.

Vikao software is for Palm handhelds and it allows you to take images and screen shots from your computer and transfer them to your Palm device as an image. This can be a great way to capture an Internet map to take with you on your device.

This software can be downloaded at no cost from **http://www.mayeticvillage.com/**

In practice

At Bishop Burton College, PDAs have been used for fieldwork to identify weeds. The tutor added re-sized JPEG images of weeds onto the devices and the students used the HP Image Viewer to compare and identify what was seen in the field with the gallery of images on the device. Written notes could have been taken to name and label the weeds in particular. 'Students could easily see the images despite the tasks being undertaken on a sunny day and had little difficulty in navigating from one image to another. They found the task much more interesting than identifying weeds from a book or from copied drawings.' **(http://www.adveb.co.uk/mlearn/pdaresea.doc)**

John Whalley and Philip Percival

Capturing images

Images can be captured on many mobile devices. This can be used for learning in a number of ways including:

> providing evidence of a learner performing a physical skill, e.g. yoga position;

> providing evidence of an artefact created, e.g. a clay pot;

> illustrating a piece of written work or an oral presentation;

> recording stages during a process of creating an artefact.

Taking photos is a great way to capture an event as it is happening, whether in a classroom or outdoors, and can provide a stimulus for later reflection on the experience.

Peterborough Adult Learning Service use digital cameras and MP3 players to take pictorial and sound recordings to capture progression and achievement in pottery classes.

In practice

'My son doesn't eat well which is difficult when we take him out to restaurants. The PDA is great as we can photo him as he eats a new food. He can write down or draw what he's eaten. It's a good motivator.'

Parent, Bilston Nursery

Using Short Messaging Service (SMS) and Multimedia Messaging Service (MMS)

Short messages can be a very effective way for learners to communicate swiftly and succinctly with their peers and with their tutor.

Many mobile devices are capable of sending or receiving messages. This feature can be used for learning in a number of ways including:

- text communication between peers for opinion sharing, feedback or to request support;
- capturing reflections and thoughts on a weblog using text and/or images;
- providing interactive quizzes which can offer SMS feedback;
- creating a 'treasure hunt' activity capturing photos and text whilst 'out and about' using a web-based storage area or moblog;
- sharing images of a personal project with peers;
- capturing images and video clips to enable the learner to record their progress, and then share the files with peers and tutors.

Sending multiple messages

Some organisations provide a service so that tutors can manage, send and receive SMS text messages from a desktop computer with Internet connection. Messages can be sent quickly to individuals and groups. They can either be sent immediately or scheduled for delivery at certain intervals. Although this type of service is not usually free, the advantage is in the tracking of message delivery and the control of scheduled messages.

One example of a provider offering an 'education messaging community' is Txttools.co.uk. With this service you can manage, send and receive secure SMS text messages. Their 'Keep Warm' campaign suggests that learners can be supported from application to enrolment using regular text messages over the summer period, for example, keeping the learner advised of start dates, tutor information and equipment required. Throughout the first term the messages can offer support and guidance, advice if tutor sickness or bad weather prevents a class from starting or if the venue has changed. Learners' questions can be captured and answered via the web-based service and replied to via text message. **http://www.txttools.co.uk/**

TRY A useful e-book, *Using SMS to Access Google*, can be downloaded from the website of the JISC Regional Support Centre in the North West.
http://www.rsc-northwest.ac.uk/acl/BookCase/library_section2.htm

If you have a camera phone or a PDA/phone with an in-built camera it will have MMS or picture messaging functionality. That means you can send images to another mobile phone user who has MMS, or send the images to a web area to store and share them. Web areas like Moblog, PocketBlogger or Ploggle can receive content from your mobile phone or PDA/phone and allow you to add text annotation to each image.

Bulletin boards

Bulletin boards offer an online space where messages can be posted and read by learners. They can be used for educational discussion and for information sharing.

Bulletin board systems, such as media board, can receive SMS, MMS and email as well as web access to the board. One provider of media board space is Cambridge Training and Development (CTAD). Their media board can be set up to look like an Internet message board using a visual image as a central focus. The content – audio, text, images and video – can be sent from a computer or a mobile device and is added to and around locations within the image.

Media boards can be used for field trips so learners can explore and take photos, type text messages and send the images and text to a media board for later analysis in class.

In practice

A media board can be used interactively in learning tasks such as:

- working as a team to give and receive directions and instructions or negotiate and agree how to solve a problem;
- following directions or answering questions to complete a task;
- making enquiries, conducting interviews or surveys, and recording speech or other audio.

(http://www.m-learning.org/which.shtml)

TRY To see what a media board looks like go to **www.mboard.co.uk** and click on any media boards that do not show a 'lock' symbol.

Using the planning tools

The planning tools available on many devices can be used for learning in a number of ways including:

> to practise time management skills by creating calendar appointments;

> to practise maths skills – awareness of time, length of activity, sequence;

> to use calculators to support other maths activities;

> to practise English skills looking at the A-Z sequence of the address book;

- to work collaboratively with learners to create a group list of learning outcomes for a course/programme using memo pad/ task to do list;

- to set alarms and timers for group activities (for example, length of a discussion or activity).

For tutors it is important to be aware that knowledge and understanding of basic maths principles are required to practise calendar and calculations including the 24-hour format of time.

All PDAs and many mobile phones come with some form of personal information management (PIM) software. This can include:

- Calendar
- Clock
- Alarm
- Notes
- Tasks/to do list
- Contacts and address book

Some PIM functions include additional tools like currency converters or calculators.

Redcar Adult Learning have trialled the use of Personal Digital Assistants (PDAs) within their management team: 'The time saved in not having to type up my notes after an observation is great. A real saving!' The use of a PDA is seen as unobtrusive and as it's quick to turn on there is no delay in capturing the vital evidence during observations.

Cornwall Adult Education Service wanted to achieve a more efficient way of collecting essential data and therefore developed a Personal Learning Record, which was designed to be accessed on a PDA. This has helped the learners by making the form easy to fill in at anytime throughout the session and has reduced the amount of paperwork to be completed. Although some learners indicate they prefer the 'safety in paper', they appreciate the easy and private method to complete the Personal Learning Record and at the 'touch of a button' upload the form to their own online learning area on the learning environment – Moodle. The completed form is updated in real-time using a wireless Internet connection at learning centres. This has helped speed up the collation of information and has allowed easy access to review the document as a live part of the learner's programme.

In practice

'My FiloFax is now redundant. I make notes using the onscreen keyboard, as I can't read my writing on the screen. It doesn't like Copperplate writing. The PDA is powerful as a diary and aide-memoire. My time management has improved – with alarms set to tell me my next appointment. Fundamentally I use it as a diary but it's also brilliant to have a calculator. I have also used it as a Dictaphone.'

In practice

'The learners' Individual Learning Plans (ILPs) on the PDA need to be private and confidential. I supplied SD (Secure Digital) cards to each of the learners. I can then pop my SD card out before lending the learner my device.'

Skills for Life tutor, Cornwall Adult Education Service

Doncaster Metropolitan Borough Council uses Blackberry 7290 devices to keep adult learning tutors informed of training events, minutes of weekly meetings and other useful and relevant information, and it's a way for the tutors to advise of learner attendance, queries or requests for resources. The tutors can also support their learners using the email function. The Outlook calendar shows everyone's commitments and appointments and a booking system, using the calendar, allows tutors to reserve equipment.

Wolverhampton Adult Learning has encouraged learners on a parenting course at Bilston Nursery School to use a PDA to capture details of their child's development away from the school. They use management functions to add contact numbers to the Address Book, add doctor and dentist appointments into the diary and to take notes. Many also use the alarm and timer function during cooking as well as the calculator to check bills and prices when they're shopping. As the parents have started to use the devices in their own lives, so this has increased their confidence to use the devices to capture images, sound and video to support their child's development and support their own learning.

TRY Have a look on your own device, or one you can borrow, and find the calendar function and enter in an activity for tomorrow.

Did you know...?

Devices like the LG Electronics F7100 Qiblah phone have an in-built reminder system for Muslims to pray (Azan time) and a compass to indicate direction.

Did you know...?

The 5th Generation iPod (Video iPod) not only plays video and audio files, it also allows the storage and viewing of photos in a gallery, viewing PowerPoint presentations, and offers simple PIM functions to let the user set calendar reminders, alarms, take notes and also play games.

Quizzes, games and bespoke resources

Quizzes, games and bespoke resources can be used on many mobile devices. They can be used for learning in a number of ways and these include:

> quizzes to review or 'drill and practice' activities to re-visit new learning;

> games to present content in an engaging, enjoyable way;

> collaborative learning such as working in pairs to 'play a game';

> bite-sized chunks of learning to extend learners who have completed a class activity;

> reviewing activities to help learners who need extra support;

> encouraging learners to design their own resources.

PocketExam, Quiz Buddy and Quia are all software applications which offer you the opportunity to create your own quizzes using multiple choice, flash cards, yes/no or single choice activities including image prompts or sound files. Microsoft Excel can also be used to create interactive quizzes.

Quizzes using SMS

Text messaging (SMS) can be used by learners to submit answers to exercises and obtain immediate feedback.

In Buckinghamshire Adult Learning, *Skills for Life* tutors have created interactive quizzes to engage learners who are reluctant to use technology. Each learner or group works through a set of questions and then uses the PDA to send the answers by SMS to a dedicated phone number. The learner/group then automatically receives the score and feedback by SMS. Groups of learners found this activity engaging and one said that it was a 'cool way to learn'.

One tutor engaged a group using visual prompts and quite complex maths questions. An example of his quiz paper is shown below. The tutor found that this group activity formed the basis of a sustained episode of learning.

Worksheet

HD1/L12.3 M-learning- SMS quiz – (sms-pro-01) tutor D Hill

Complete quiz then text the word pro (followed by the five numbers of your answers) to 07800*****

Name:	Date:
A) What is the mode of holdings - % net assets 30/06/04?	
1	2.7
2	2.4
3	2.6
4	2.45
B) What is the median of holdings - % net assets 30/06/05?	
1	2.85
2	2.78
3	2.95
4	2.9

Example of worksheet used by Buckinghamshire CC

You can find out more about SMS quizzes and the cost of the service from CTAD at **http://www.m-learning.org**

Learning games

The BBC has a range of purpose-built quizzes and games for mobile phones in the Bitesize section of the BBC website.

The Food Standards Agency has a section of its website available for PDA and mobile phones so you can receive food safety and nutrition advice when you're out and about. There is a breakfast and barbeque game, a GM and food labels quiz, as well as healthy calculators. See **http://www.eatwell.gov.uk/**

Downloadable games such as Scrabble and Suduko can offer an opportunity to practise literacy or numeracy.

These activities can be used as a group activity where the results of each move can be 'beamed' to a partner's device so each 'player' can take turns.

In practice

'The inbuilt game "Jawbreaker" is really useful for all the new Ipaq PDA users to become familiar with using the stylus. The bubbles or balls have to be "popped" by tapping onto the screen and you can score. It has become quite a competition.'

Manager, Redcar Adult Learning

Interactive resources

CTAD (Cambridge Training and Development) has researched, developed and trialled the use of its m-learning materials for Pocket PC. To find out more about the project background, and to see the authoring tool which will allow a user to create bespoke resources for the Pocket PC, go to **http://www.m-learning.org/**

The health and fitness 'Know your Food' m-learning materials have been used by Gloucestershire College of Art and Technology

(GLOSCAT). GLOSCAT runs Healthy Cooking courses at the request of centres concerned about the diets of children in Early Years/School settings. The learners engage in literacy, language and numeracy activities as well as cooking healthily. They use the CTAD m-learning materials as a warm up before the class starts.

In practice

'I'm trying to combine cookery and basic skills so I want to encourage the group to use the PDAs. The classes are usually busy and I see them being used by learners at different times. Handing someone a PDA when they have five minutes to spare for example or as people arrive or while others are washing up etc. They can use it to:

- View short 30 second tips taken from www.eatwell.gov.uk
- Work through the m-learning quizzes
- View a Powerpoint presentation from the British Nutrition Foundation'

Family Learning tutor, GLOSCAT

In practice

'Traditionally you're supposed to write down the words to describe the movement, which is strange for something so practical and visual. The task was to create a stick person symbolising the exercise, cut and paste the routine so you can build up a sequence and then you have a record of it. You also have to learn cues so these can be added to the animation. These would be great for any exercise class.'

Principal and learner, Cornwall AES

TIP If you're going to make your own PDA web pages, consider the following:

- Use 2 or 3 columns maximum
- The width must be no larger than 220 pixels wide
- Use Arial, size 10, black for the 'body' text attributes
- Use colour as much as possible
- Make sure any images are no bigger than 220 pixels wide
- Why not add animation for fun?

Try the website **http://webdesign.about.com/od/pdas/a/aa060500a.htm** for helpful tips and hints on creating web pages for PDAs.

Concept mapping is a visual technique for working with ideas, similar to flow charts. Learners could work collaboratively on a concept map sharing and encouraging peer critique. Concept maps for PDAs include Pocket MindMap, PicoMap, and IdeaPad for Palm.

There is a wide range of downloadable programmes for Pocket PC or Palm PDAs and an increasing number for the RIM Blackberry devices. Some are free and easy to download, and synchronise to your device with little or no difficulty.

TRY For free downloads use a search engine like **http://www.google.com** and search using the keywords 'free downloads for Pocket PC'. Alternatively replace the keyword for 'Palm'. Many downloads are free or offer a free trial.

Checklist to help you identify the functions and facilities of a handheld device

Do I require this functionality?			How could I use this in my teaching?
	YES	NO	
Voice call/phone facilities			
Web browser			
Email / POP3 setup			
PIM functions:			
Calendar			
To do/task list			
Clock and alarm			
Contacts list or address book			
Notes			
Converter			
Calculator			
Text reference:			
E-Book reader e.g. Microsoft Reader			
Document viewer e.g. Mobile Word, Pocket Excel			
PDF reader			
Audio reference:			
Sound recorder			
Music player			
Earphone jack points			
FM radio			
Image reference:			
Camera			
Camera – video functionality			
Image viewer/gallery			

Do I require this functionality?			How could I use this in my teaching?
	YES	NO	
Video player			
Image mobile printing			
Projection software e.g. Clearvue			
Flash player			
Recording:			
Handwriting recognition and recogniser			
Communication/collaborative:			
Bluetooth enabled			
Infrared enabled			
SMS functionality			
MMS functionality			
GPRS			
Wireless or WIFI			
Instant messaging e.g. MSN Messenger			
Touch screen			
Stylus			
QWERTY keyboard – on-screen			
QWERTY keyboard – on-device			
Phone keypad			
Synchronisation docking cradle			
Synchronisation cable			
Flash Data Card slot			

4

Out and about and getting connected

Out and about

The portability of handheld devices is particularly useful to engage learners who prefer to be active and outside. The examples below show two ways of producing a guidebook for a local area using mobile technology.

Example 1

Images of location and background information can be stored on the device prior to a visit and learners can add text on site during the visit.

Example 2

Learners can use the device to take photos, write text notes or make sound recordings during the visit. The files can then be collated later at a computer or beamed to other devices to share and work collaboratively to produce a final guidebook.

The Global Positioning System satellite navigation system (GPS) provides unique location information. GPS is used in marine navigation, by hill walkers and in car-based systems. Many mobile phones can utilise this GPS 'positional awareness' technology. The technology now allows an 'augmented reality' where buildings or objects can 'talk' and offer information about themselves to PDA or mobile phone users. This information can also be tailored to the user's needs, taking account of level of language or preferred format.

In practice

At Bishop Burton College, learning activities took place in field and farm environments as well as the classroom. The PDAs had pre-prepared Excel spreadsheets and databases designed for the devices, allowing students to collect and analyse data in a 'live' environment.

Trial projects have identified that PDAs are useful in science fieldwork where the compact size of the device has a great advantage over bulkier laptops. Spreadsheet software can cope well with data input and analysis. Word processing software allows for note taking and it is possible to connect the devices to other measuring and sensor equipment. Opportunities for audio recording and digital images enable learners with disabilities to have more options for recording fieldwork experiences.

For instance, if you are an amateur naturalist you might enjoy going for a walk in the countryside. Should you hear a particular bird and cannot recognise it you could use your PDA or mobile phone to access **http://www.rspb.org** and check the bird sound, see a picture of the bird and read information about its habitat and other useful information.

Oxford Brookes University scientists designed Wildkey interactive software with the developers Adit Limited to revive interest in biology-based fieldwork. Wildkey was produced for biological identification and recording for handheld devices using Pocket PC or Mobile Windows. The software enables learners to identify species in the field through simple questions and images. Data that has been recorded can then be investigated in map or graph form on a handheld device to identify patterns in species distribution and behaviour.

TRY For outdoor work bear in mind that some screens will be difficult to read in bright sunshine. Transflective or TFT screen types show good resolution in both reduced light and in bright sunlight.

In practice

'Some work with GPS-enabled PDAs has helped young people with learning difficulties understand map reading. When they can see their position plotted in real time on the map it becomes easier to relate the map to the features around them. The screen shot shows in a thin red line the learner's route, along with the doubling back that took place when they self-corrected the route. The instant feedback and the encouragement to self-correct is very valuable for this group.'

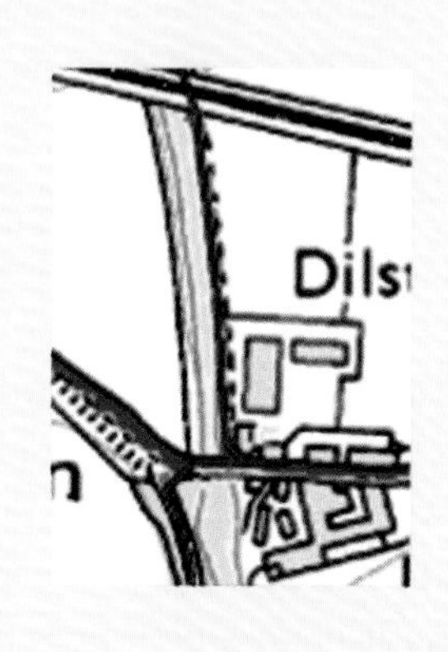

Alistair McNaught, Senior Adviser, TechDis

This type of activity not only improves map reading and orientation but can also encourage team work and develop communication skills.

Getting connected

Many mobile devices can connect to the World Wide Web through the Internet, but may also be connected to a single device in close proximity through 'beaming' technology.

The Internet and beaming can be used for learning in a number of ways including:

> Each learner can access his or her own specific materials.

> Learners can interact with the tutor and other members of the group outside the classroom.

> 'Beaming' between devices can encourage interaction and collaboration.

> Beaming' between devices can encourage learners to interact. Anonymous questions can be received by the tutor via a 'beam' from learners.

Beaming

The screen of a Pocket PC showing connection settings

Bluetooth and infrared technology allow you to 'beam' files to another person's device. Beaming allows simultaneous interactions between people and it can encourage better thinking and problem-solving skills as learners can work together on a project, sharing images, sounds and data between devices.

A storyline could be started by one learner, the document 'beamed' to another for the story to be added to, then passed to the next learner.

In practice

'A *Skills for Life* group has been using the voice recorder as part of the Communications module. They create a message and beam it across to another device as a 'chinese whispers' activity to see if the original message remains or whether miscommunication has occurred during transit and beaming. They really liked the activity so I will plan to include more beaming in future.'

Skills for Life tutor, Redcar Adult Learning

Beaming can be achieved using Bluetooth or infrared technologies. Bluetooth has a range of 10 metres and once the function is turned on, the device *'seeks'* other devices such as phones, printers and cameras in its neighbourhood. Infrared has a shorter range of only one metre and ideally your device needs to be immediately next to the infrared *'window'* of the communicating device. If many devices are connected by Bluetooth in the same location then the transfer of files and information becomes slow. See **http://www.bluetooth.com**

TIP When you turn on Bluetooth on your mobile device always remember to turn it off after the activity to prevent a drain of power on the battery.

The Internet

When you're on the move it can be very useful to be able to connect to the Internet and access learning material or an interactive web-based quiz. Once connected, it can bring the web experience to the handheld device user.

In practice

'Learners who are blind and also not familiar with computers can find it a daunting experience to learn to use a screen reader in order to understand how to access the Internet. Harrow Adult and Community Learning has run a project trialling the use of speech-enabled smartphones to search the Internet and to send and receive SMS messages'.

Project Manager, Harrow Adult and Community Learning

'This training has changed my life in so far as I am better able to communicate with the sighted world as a visually impaired person.'
Learner, Harrow Adult and Community Learning

Cornwall Adult Education Service regularly uses PDAs in *Skills for Life* classes. Although the learners have access to many computers in the IT suite, they choose to use the PDAs when working at a paper-based activity for privacy and ease. They connect to the Internet and use **http://www.dictionary.com** to check spelling and grammar or to search for resources generally.

Another Cornwall tutor works with a group of learners at a travellers' site and she finds carrying the small devices much more convenient than taking in quantities of laptops. Electricity is sometimes an issue so PDAs can be used 'away from the mains'. The learners enjoy accessing the Internet – for many it's their first experience of using the Internet and sending emails.

In practice

Doncaster Metropolitan Borough Council uses its Blackberry devices regularly to access the Internet for resources on **http://www.studystack.com**. 'This is a good site offering Flashcards, Matching exercises, Crosswords. There are even Hangman games. They work well even with a small screen.'

Tutor, Doncaster MBC

Many PDAs don't easily connect to the Internet and the choice and method depends on the device and its functionality. Also, the connection speed can be slow and the web pages difficult to see without copious scrolling. Some websites now indicate whether a PDA/mobile equivalent website is available and Google can help you easily search for PDA appropriate sites.
http://www.google.com/mobile/

TIP

For a wide range of technical advice and information see BECTA's Technical Paper entitled *Mobile Internet Connectivity.*

Did you know?

AVANTGO software allows you to download selected web pages during synchronisation for offline reading. This can allow you to select information to read at a later date. **http://www.avantgo.com**

5

Are these devices accessible for everyone?

Mobile learning has the potential to offer the advantages of e-learning to a wider group of learners in a range of environments.

As with any teaching and learning approach, there are both accessibility constraints and opportunities. The challenge for the tutor is to grasp the opportunities afforded and match them to the needs of the learner where value can be added to the learning experience. It is important to use technology to remove barriers for learners and recognise that, for others, they may create them. NIACE has worked closely with the JISC TechDis advisory service to develop pragmatic guidance on the accessibility of mobile technologies.

This section offers you a model to use in planning the use of handheld devices in a learning activity. The model helps to identify how accessible the experience is for each learner.

Michael's experience

Let's meet a learner and find out how accessible he finds the experience of using a handheld device in his learning.

Michael has learning difficulties and is partially sighted. He has recently moved into his own accommodation and wants to increase his independent living skills. The local college runs an adult 'Cooking for Beginners' class for people with learning difficulties and the tutor has been experimenting with creating podcasts of recipes. Michael's vision means his MP3 player is difficult to use due to the small screen. The cookery tutor has recorded different recipes for each week based on the skills they learned in the class.

Michael finds this a huge aid to his independent living. He takes a long time to find the right track on the MP3 player but, once he has, he can stop, rewind and play the recipe much more easily than he can use a recipe book. He loves telling his friends and carers about his 'podcast cooking' and explaining to them how it works.

CONSIDER What could the tutor do to help him? Suggest two ways to improve Michael's experience.

The M-Learning Accessibility Model

Let's look at what accessibility means and then consider a model to see how the tutor could help Michael and other learners. The model below has been devised by the organisation called TechDis. You can use the axes to plot your position and indicate where any use of mobile technology will meet users' needs.

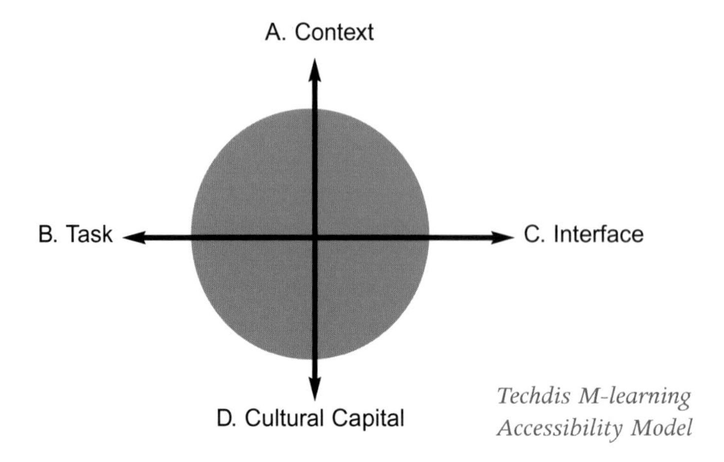

Techdis M-learning Accessibility Model

A. Content	=	Does the content on the device support me?
B. Task	=	Does the task engage and motivate me?
C. Interface	=	Can I see the screen well? Can I work it or hear it?
D. Cultural Capital	=	What value do I put on the experience using this device?

Putting the model into action

Imagine you've been given a set of PDAs to use in a session. Think of a group of learners and a suitable activity.

Step 1. Take one learner and plot the reaction to A,B,C and D. How do you think s/he would respond? 0 (zero) is 'little' and at the centre point, 5 (five) is 'greatly' and is at the end of the arrow point.

Step 2. Mark an x at each of the four points relating to your response.

Step 3. Join up the four points to form a diamond/square shape.

So what does that mean?

A diamond/square shape located close to the centre point of the model, or within the coloured area, indicates that the activity is *less* accessible for that learner. If this is the case you will need to consider an alternative activity or method.

We return to Michael and see how the model can be applied to his situation.

His application of technology could look like this:

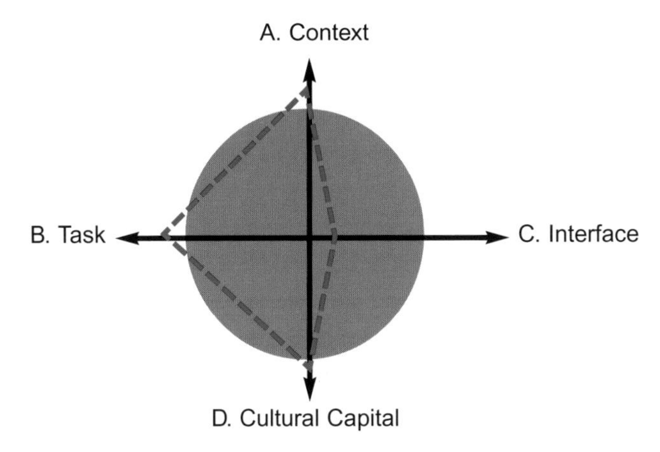

Accessibility could have been improved for Michael by using a clip-on magnifier such as the Magnifico.

Deepa's story

Deepa attends an English class where the tutor loans PDAs to the learners to make notes of shop names in the local centre and take photos of the shops. In the class session they work in small groups, beaming the information to one another and the tutor. The tutor then uses presentation software to talk through each group's photos and develop language around them. Deepa is reasonably competent to work the interface and finds the language development really helpful, but she is not comfortable for cultural reasons because the group work involves talking to men and she does not like 'gadgets'.

Deepa's model could look like this:

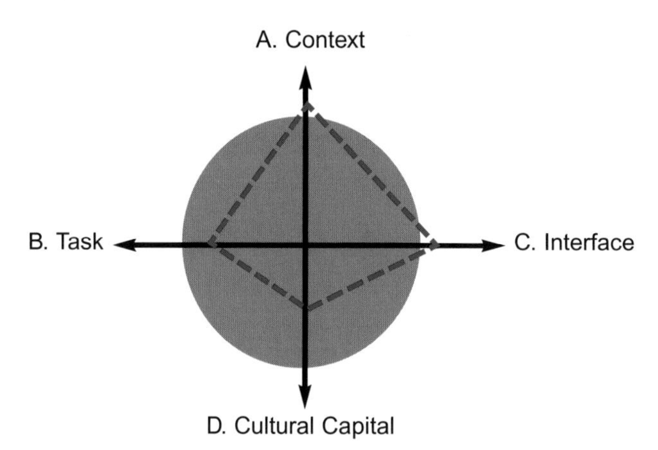

Accessibility could have been improved for Deepa by:

- Reorganising the group she is working in.
- Providing her with a paper notebook so she only used the technology as a camera rather than note taker.
- Explaining that the PDAs were only being used for this exercise and other methods would be used later.

Why not ask the learner you considered earlier whether he/she agrees with the model you made. Identify ways to improve the accessibility of the experience. List the suggestions and decide how you can put them into action.

	Suggestion	Action

For an additional learner's story and a link to this model see **http://www.niace.org.uk/mobiletechnology**

Learner centred accessibility

Mobile technology may be a more portable and affordable way of bringing the benefits of e-learning to classrooms, community halls and other technology-challenged environments. From the examples above it is clear that accessibility needs to be seen in relation to:

- the whole learner experience – what they do and how they do it;
- the alternative learner experiences – what they would otherwise do to achieve the same learning objectives;
- the alternative resources – whether mobile learning offers more flexibility than traditional resources, such as handouts.

TIP For details on a range of mobile learning approaches and their accessibility pros and cons see the m-learning section of the TechDis website. This includes a wealth of ideas to try with learners and some 'How to' articles to get you started. **http://www.techdis.ac.uk/**

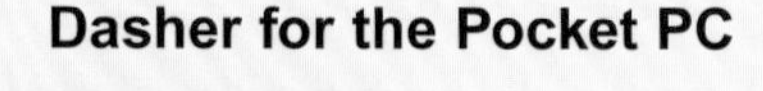

Dasher for the Pocket PC

Dasher is free software which allows a different way of writing.This is one of several pieces of software which can enable word prediction.

http://www.inference.phy.cam.ac.uk/dasher/

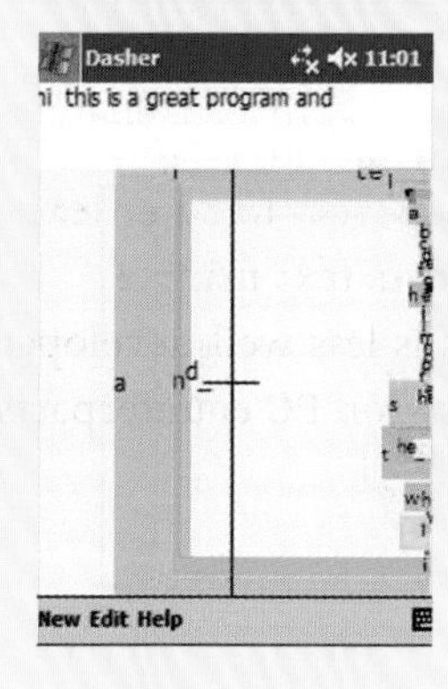

Accessibility – key themes

Key accessibility themes to be aware of when working with mobile devices include:

> **Changes to the user interface** – You can make the font easier to read by changing the font size or style. You could also clip an external magnifier to the device.

> **Alternative text entry options** – Learners can be offered different ways to enter text to take account of learners' preferences and needs. Most devices allow entry via a portable keyboard. Some offer inbuilt keyboards or handwriting recognition and

software is available to allow a wide range of alternative text entry possibilities.

- **Mobile devices support reading or referencing** – Text size, font style and colour combinations can be very important to ensure that written material can be accessed by learners. Reference texts on mobile devices can be personalised via software that allows bookmarking, adding notes, changing text size/contrast and even autoscrolling.
- **Word prediction** – Word prediction can be useful for learners who normally enter text slowly. This facility can be enabled on a number of mobile devices.
- **Time management and organisation** – Tools to support this are inbuilt to most mobile devices and can benefit learners with short-term memory or organisation difficulties.
- **Voice recognition and text-to-speech** – Where learners are blind or have restricted use of their hands they may wish to enter text using speech recognition software and to access text using a 'screen reader'. The software to support this is less well developed for mobile devices than their higher specification PC counterparts. However, systems do exist such as MobiSpeak.

6

How do I get started?

The pedagogy should always drive the technology and not the other way around. Make sure that all your decisions lead you to offer improved and more varied learning experiences.

In exploring the application of mobile technology to your learners you should start by identifying the areas of the learning experience which could be improved and then seek ways in which the technology may be able to provide solutions.

Consider the checklist in Chapter 3. Select three areas where you would like to improve your learners' experience and then suggest possible uses of mobile technology.

	Element of your learners' experience which you would like to improve	1. Possible applications of mobile technology to address this	2.. What will you do now? e.g. experiment with technology, research case studies of similar use etc.
1			
2			
3			

Would the technology enable new kinds of thinking in your teaching?

Bloom's Taxonomy is a reference model for those involved in teaching, training and learning and reflects different forms and levels of thinking. The Taxonomy provides a useful checklist to start planning activities and delivery methods to encourage deeper levels of thinking amongst your learners. It can also be a tool to use when deciding on the next step you take with mobile technology.

Competence	Skills demonstrated	Examples of activities with handheld devices
Evaluation	The learner makes decisions based on in-depth reflection, criticism and assessment.	Mobile blogs
Synthesis	The learner works collaboratively to consider various points of view and makes links to theory.	SMS collaboration, moblogs, bulletin boards systems, media board
Analysis	The learner breaks learned information into its parts to best understand that information.	Personal notes, Web searching, simulations
Application	The learner makes use of information in a context different from the one in which it was learned.	M-Learning materials, GPS, photo gallery
Comprehension	The learner grasps the meaning of information by interpreting and translating what has been learned.	SMS and general quizzes, beaming
Knowledge	The learner is able to recall, restate and remember learned information.	Reference materials, notes, podcasts, e-books, image gallery

Benjamin Bloom's Taxonomy
Source: Bloom (1956)

7 Glossary

3G	The third generation telephone network transferring data at 384 kbps. This allows for video transmission as well as standard data transfer. See BECTA's Technical paper: *Mobile Internet Connectivity*
4G	Fourth generation telephone network is still in development. Allows for 'roaming' between networks to ensure a constant connection to Internet at all times. See BECTA's Technical paper: *Mobile Internet Connectivity*
Beaming	Method for transfering files from one device to another. See Bluetooth and Infrared.
Blackberry	Produced by Research in Motion, a Blackberry is a handheld wireless device providing e-mail, telephone, text messaging, web browsing and other wireless data access.
Bluetooth	Short-range wireless technology for transmitting voice and data across a global radio frequency band. Used to connect devices and peripherals such as mobile phones, cameras, printers

and headsets. (Transfers at 721kbps, Bluetooth 2 transfers at 2.1mbps)

Character recognition — Converts hand-drawn symbols to digital format and then to alphanumeric characters.

Docking cradle — A device used to connect a handheld device to a PC for synchronisation and application downloads. The two are connected via the handheld's communication port using a serial or USB cable. The cradle often doubles as a battery charger as well.

Expansion slot — The opening in a handheld device where standard cards can be inserted to add storage memory, network cards and wired or wireless modems, GPS units, or cameras.

GPRS — GPRS stands for General Packet Radio Service. The fast 'always on' service allows you to connect to the Internet anywhere using a wireless modem connecting at speeds of 40kbps (a dial-up modem speed). It keeps you permanently connected to the Internet and charges only when you're sending or receiving data.
See BECTA's Technical paper: *Mobile Internet Connectivity*

GPS — Global Positioning System. A satellite-based location technology that can determine position. It can be added to handheld devices.
See BECTA's Technical paper: *Mobile Internet Connectivity*

Handheld	A portable device for storing and managing personal information. The available features depend on the type of handheld you choose.
Infrared beaming	The process of using the IrDA (Infrared Data Association) standard to transmit data wirelessly from one device to another. Beaming is a way to exchange files and applications between handheld devices.
LCD screen	Stands for Liquid Crystal Display. The type of display screen on many handheld devices. See also TFT
Microsoft Origami	Microsoft's portable media player with PDA and games functionality.
MP3 player	A digital audio player. It can store, organise and play music files, not just MP3 file formats but potentially Windows Media Audio or Advanced Audio Codec. Some MP3 players can record sound in WAV file format.
Operating system (OS)	The program that is stored in the Read Only Memory (ROM) and controls a handheld's main features. Operating systems include Palm, Windows Mobile or Windows Pocket PC, Linux and Symbian.
PDA	Personal Digital Assistant. A small handheld computer.

Personal information manager (PIM)	A program for organising contacts, appointments, tasks, and notes.
Push email	A device which collects and sends email to your device automatically, rather than requesting the message.
Random access memory (RAM)	The segment of a handheld's memory that stores data and applications. In some handhelds, all data in RAM is lost when the battery power drains.
SD	Secure Digital Card is a solid-state removable memory card which is used with digital cameras and handheld devices.
Smartphone	A mobile phone with PDA functionality.
Stylus	A pen-like device for navigation and data input. Most handhelds have a slot for storing a stylus.
Symbian	Symbian operating system (formerly Psion) for PDAs and mobile phones
Synchronisation	The process of exchanging data between a handheld device and PC so that changes are reflected on files stored in both computers.
TFT	A transflective screen on a handheld device, similar to an LCD screen but uses thin film transistor technology (TFT). Copes well with bright sunlight or poor conditions when using a handheld device outdoors.

WAP	Wireless Application Protocol which enables a slimmed-down version of the Internet to appear on the screen of a mobile phone. Travel information and sports headlines can be delivered to WAP enabled devices. See BECTA's Technical paper: *Mobile Internet Connectivity*
WiFi	WiFi products operate using a wireless network based on IEEE802.11 standards. It is a faster method to connect devices than Bluetooth and allows wireless access to the Internet. See BECTA's Technical paper: *Mobile Internet Connectivity*

8

Useful websites

Anytime, Anywhere Learning – Planning for Success
http://www.microsoft.com/education/aalsupport.mspx

BBC for mobiles and PDAs
http://www.bbc.co.uk/mobile/web/index.shtml

BBC for 3G mobile phones
http://www.bbc.co.uk/mobile/web/highlights.shtml

Cambridge Training and Development (CTAD)
http://www.ctad.co.uk/

Dolphin screen reader and Pocket Hal for PDAs
http://www.dolphincomputeraccess.com

Encyclopedias
http://www.mobipocket.com

Forum
http://www.handheldlearning.co.uk/

Jon Trinder
http://www.ninelocks.com/

Language software for Pocket PC, Palm and Smartphones
http://www.lingvosoft.com

M-Learning Portal
http://www.m-learning.org/

Microsoft Reader software for Pocket PCs
http://www.microsoft.com/reader/downloads/ppc.asp

Moblog
http://moblog.co.uk/

MobileSpeak
http://www.accessableworld.com/mobile_speak/

Ploggle
http://www.ploggle.com/

Pocket Slideshow:
http://www.cnetx.com/slideshow/

Palm e-books
http://ebooks.palm.com/

PocketExam software
http://www.bizon.org/pocketexam/tour.htm

Phone monocle
http://www.magnifics.co.uk/

Quizzler software
http://www.quizzlerpro.com/

9

Further reading

Attewell, J., Da Bormida, G., Sharples, M. and Savill-Smith, C. (2005) *MLearn 2003, Learning with Mobile Devices*, LSDA, http://www.isda.org.uk/files/pdf/1421.pdf

Becta (2003) *Mobile InternetConnectivity*. Technical Paper, Becta, http://foi.becta.org.uk/content_files/corporate/resources/technology_and_education_research/mobile_internet_connectivity.pdf

Bloom (1956) *Stability and Change in Human Characteristics and Taxonomy of Educational Objectives.*

Hamill, L. and Lasen, A. (2005) *Mobile World – Past, Present and Future*, New York: Springer

JISC (2005) *Innovative Practice with e-learning*, HEFCE, http://www.jisc.ac.uk/publications/publications/pub-innovativepetaspx

Kukulska-Hulme, A. and Traxler, J. (2005) *Mobile Learning – A Handbook for Educators and Trainers*, London: Routledge

Prensky, M. (2001) *Digital Natives, Digital Immigrants.* http://www.marcprensky.com

Wagner, E.D. (2005) 'Enabling Mobile Learning', *EDUCAUSE Review*, May/June 2005 http://www.educause.edu/ir/library/pdf/erm0532.pdf